# UoLearn
## Easy 4 me 2 learn

Other books from UoLearn

Order books from your favorite bookseller or direct from www.uolearn.com

# Report writing skills training course.

## How to write a report and executive summary, and plan, design, and present your report.
## An easy format for writing business reports.

## Lots of exercises and free downloadable workbook.

Published by: Universe of Learning Ltd, reg number 6485477, Lancashire, UK
www.UoLearn.com, support@UoLearn.com

First Published 2010

Reprinted 2012

ISBN  978-1-84937-036-3

Other editions:  ebook pdf format 978-1-84937-046-2

Other imprints:
Easy 4 Me 2 Learn How to write reports in easy stages. How to write executive summaries, plan, design, structure and present your report. 978-1-84937-035-6

Skills Training Course, Universe of Learning and UoLearn are trademarks of Universe of Learning Ltd.

Photographs © www.fotolia.com
Cover photo © Yuri Arcurs, www.fotolia.com

# Praise for Report Writing

*"Very useful, it has given me a great number of useful tips and information."*

*"This will help me to organize and structure my work in a logical way."*

*"I will now have a structure to use when preparing future reports and I will feel much more confident in doing this."*

*"I now understand how important preparation is when writing a report."*

*"The list of eight questions is fantastic, I will always use this for my reports from now on."*

*"Very informative and useful."*

*"These techniques can be applied in my work straight away."*

*"Extremely interesting and useful – vital preparation if you need to write reports for work."*

*"Very easy to take back into the work environment."*

*"Practical advice and guidance plus plenty of materials to support my report writing."*

# Dedication

I would like to say thank you to all five of my parents, my first ones who have been with me throughout Christine and Ken Vaughan, my extra one whilst growing up Margaret Jones and my two that I was lucky enough to share when I got married, Irene and Gordon Greenhall. They have all helped me in so many ways and I have grown in their love and support.

I'd like to say a massive thank you to Martin, my husband, who has supported me through so many ventures and put up with my odd working habits. I love you very much.

Finally a thank you to Sally, my unofficial mentor, without whom this particular book wouldn't have been written as she was willing to let me work with her staff to help train them in report writing skills.

Thank you all,

Love

Margaret

# About the Author:
# Dr Margaret Greenhall

Margaret was a chemistry lecturer for eight years and it was during this time that she was asked to teach the foundation study skills to the new students. This started her interest in how people learn and how the learning environment can help them learn better. She also was involved in educating adults through the access to science course and a special project for the millennium where she helped people discover an interest in science by wading in a river and catching plastic ducks. She moved to staff development and again learned more about how people share information with each other.

She then moved to an administrative role at the University of Manchester and was suddenly confronted with tasks that she'd never done before. One of these was to write reports that went to the highest level of the university, the vice chancellors. This was scary, people assumed that coming from a highly professional background that she would know what to do. This got her investigating how to write reports and she realized that this connected with many of the skills she'd worked on with her students about taking in and sharing information.

In 2003 Margaret left the university to start a training business, specializing in helping people to understand and share information easily and efficiently. This includes topics such as speed reading, improve your memory, creative problem solving and report writing. This book on report writing combines and shares the unique style of creative planning and sharing information. Margaret can be contacted at www.inspirachange. co.uk, trainer@inspirachange.co.uk

# Executive Summary for
# Report Writing Skills Training Course

➤ **The aim of this book**

Is to teach people, with various levels of experience, how to plan, write and present information as reports. The book includes a range of opportunities to practice and develop their skills.

➤ **Excellent reports**

An excellent report is a concise and accurate record of information that contains only the data relevant to the readership and record keeping process. It will be neatly laid out and easy to read with a simple structure that allows easy access to the information.

➤ **Setting objectives**

Before you start to write your report you need to work through a series of 8 questions and write a clear objective for the report. This should then be checked by the person who initiated the report.

➤ **Planning reports**

Before you start to write the report you need to create a plan for the sections. The suggested technique is to separate this out into steps, so that you can think about what the information is and then organize it. You then need to gather all your source material together and advice is given on how to read a large amount of material.

➤ **Types of reports**

There are three main types of report

1. Information
2. Research
3. Proposal

Each has different possible sections that could be included. All benefit from a title page, aim, contents, executive summary and bibliography. In addition the components in the following table may be added. Some thought needs to be given as to whether all sections are necessary and what their order should be.

Section headings should be meaningful to your audience, not just single words.

| Information | Research | Proposals |
|---|---|---|
| Introduction | Introduction | Position |
| Categories | Method | Problem |
| Recommendations | Results | Possibilities |
| | Conclusion/ Discussion | Proposal |

➢ **Executive summaries**

A summary of the report can be included at the start, this should be no more than 5% of the report length. It should be in the same order as the report and should only include material that is in the main report.

➢ **Organizing the report**

To make things easy for people to remember, have white space, give no more than 4 ideas at once and people tend to remember starts, ends and anything unusual.

➢ **Layout style**

The recommended font style is sans serif (straight) and 12 point. People are advised to use the style formatting capabilities of their software as it makes both consistency of style and generation of contents pages easier.

There are also additional materials on grammar and diagrams on www.UoLearn.com

➢ **Editing reports**

Editing is about checking that the content matches the original objective and making the report as concise and easy to read as possible.

➢ **Proofreading**

Proofreading is about looking for errors such as grammar and spelling mistakes. It should not involve changes to the content.

➢ **Printing and distribution**

Time needs to be left before the deadline to make sure that the printers can finish the printing. A distribution list should be made to ensure everyone who needs one gets a copy.

# Flow chart for writing a report

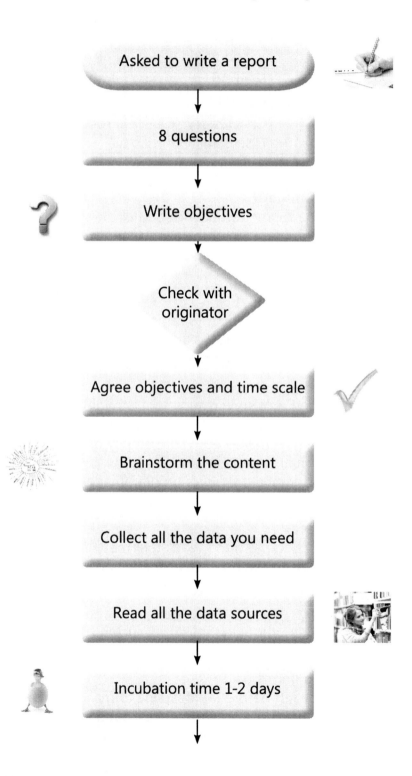

Asked to write a report

↓

8 questions

↓

Write objectives

↓

Check with originator

↓

Agree objectives and time scale

↓

Brainstorm the content

↓

Collect all the data you need

↓

Read all the data sources

↓

Incubation time 1-2 days

↓

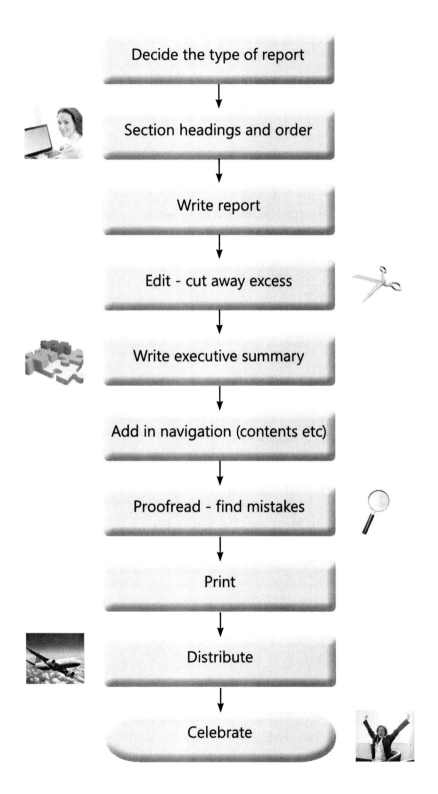

Decide the type of report

↓

Section headings and order

↓

Write report

↓

Edit - cut away excess

↓

Write executive summary

↓

Add in navigation (contents etc)

↓

Proofread - find mistakes

↓

Print

↓

Distribute

↓

Celebrate

# The aim of this book

Is to teach people, with various levels of experience, how to plan, write and present information as reports. The book includes a range of opportunities to practice and develop their skills.

# Contents

# Chapter 1:
# What makes an
# excellent report?

"I try to leave out the parts that people skip."
Elmore Leonard

# Chapter 1: What makes an excellent report?

To think about an excellent report we need to start at the question: What's the use of reports?

The word report originates from the word 'port' which means to carry and 're' which means again so literally carry again. It was used in a military sense, the report was carried from the fighting army back to the leaders to help them make decisions about the next step.

The modern definition is:

*"To bring back as an answer news or account of something, to give an account of, especially a formal, official or requested account."*
*(Chambers dictionary)*

Reports in the modern workplace have two main purposes:

➢ To share information with other people.

➢ To keep as a record of events and decisions.

Exercise: Think about reports in your work place, what purpose were they written for?

..........................................................................................

..........................................................................................

..........................................................................................

..........................................................................................

## Components of an excellent report

Collect together as many reports, either internal or external, as you can find. Look through them quickly, particularly at the layout and section headings.

What could be improved?

What features do you like?
Do they make them easier to read?

Exercise: Thinking about the 2 main uses of reports have a go at a reverse brain storm.

List 15 things you could do to write the world's worst report.

*"This report, by its very length defends itself against the risk of being read."* Winston Churchill

Here are some of the ideas from a group of participants at one of my training sessions:

**How to write the world's worst report:**

- ✘ Too much information
- ✘ Black and white - no color
- ✘ No structure
- ✘ Too many technical details no one can understand

- ✘ Not enough information
- ✘ Use really long words no one can understand
- ✘ No conclusion
- ✘ Write it on toilet paper

- ✘ No clear aim to the report
- ✘ Not enough paragraphs and bullet points
- ✘ Report on something everyone knows about
- ✘ Badly spelled

- ✘ Boring
- ✘ No date or name on it
- ✘ No summary
- ✘ Text too small

- ✘ Weighs 10 pounds
- ✘ Not sure who you're writing for
- ✘ Makes rude comments about the company
- ✘ Write it as a rap song

The truth about reports is very few people are sitting rubbing their hands with glee saying, "Goody, I've got a report to read today." A lot of the art of report writing is avoiding the 'GROAN FACTOR'. The groan factor is that sigh that we all make when we are faced with a task that we know we have to do but it is going to be boring and take a lot of our time. The way to prevent it is not to think about writing a report but to think about sharing and recording information as your goal. The main way to avoid the groan is to make the report short and to the point. Everything that should be included is there but nothing more and the structure makes it easy to read and to find information.

*"It's all about the reader." Margaret Greenhall*

---

**Definition of an excellent report:**

An excellent report is a concise and accurate record of information that contains only the data relevant to the readership and record keeping process. It will be neatly laid out and easy to read with a simple structure that allows easy access to the information.

---

Note this doesn't include any constraints on the length. A report could be a simple half page email to relay a summary of a meeting or it could be a 200 page end of project report. Nor does it have to be a paper based report, it could be a website or electronic report, even a video.

# Objectives for reports

The best method for ensuring you write an excellent report is to have objectives written down and agreed with whoever has requested the report.

The way to do this is to work thorough a series of questions, think about the answers then write an objective that summarizes your thoughts. Many people start by gathering the information and then just sit down to write and this is where you end up with rambling reports that have too much extraneous information in them and are perhaps missing vital points.

So the questions you need to consider are:

> ✓ Why is this report being written?
>
> ✓ Who is going to read the report?
>
> ✓ Who else will read it?
>
> ✓ Why do they need it?
>
>
> ✓ What do they know already?
>
> ✓ What do they need to know about the topic?
>
> ✓ What don't they need to know?
>
> ✓ What are they going to use the information for?

Let's just think about one of these questions:

Who is going to read the report?

**Exercise: Who is the report for?**

Imagine you have been asked to write a report about your department. It was a request from your head of department that you are expected to get on with on your own initiative.

What would it include if you wrote separate reports for the following different readers?

How would it change the length and structure of the report?

Your departmental colleagues.

A colleague from a different department of your company.

A group of visiting 17 year olds.

Your managing director.

External quality auditors.

What if you went back to your boss for clarification of the readership and their reply was,
"All of them, of course."?

As you can see each audience will need a different report even if you started with the same information.

The best answer to the final question, that one of my trainees came up with, was that you should have a simple short report with a website to back up the report and then within that website have different sections for each type of person.

Just one question can give great insights into the content and style of your report. So, what you need to do now is have a go at working through the eight questions with a real report that you are going to have to write. If you don't have a current one then work through the questions with a pretend one - maybe a report on your role for your manager. You can download a worksheet, with these questions, from our website (www.uolearn. com) to use when you are given a new report to write.

It is important you pick a report to work on as it will help you see how to apply the tools as you work through the book.

**Exercise: Work now on a real report you are going to write. Go through the following eight questions really thinking about the use of the report.**

Why is this report being written?

Who is going to read the report?

Who else will read it?

Why do they need it?

What do they know already?

What do they need to know about the topic?

What don't they need to know?

What are they going to use the information for?

**Some other questions to think about are:**

➤ What's the deadline?
➤ What is the lifespan of the report?
➤ How often are similar reports written?
➤ When was the last one?
➤ What version of the report is it?

➤ Do you need to put a distribution list in?
➤ Are there any data protection issues?
➤ Are there any confidentiality issues, will you need a confidentiality statement on relevant parts of the report?

➤ What type of report is it?
➤ If the report is going to a committee, is there a standard cover format or cover sheet?
➤ How long should the report be?

*"It takes less time to do a thing right
than to explain why you did it wrong."
Henry Wadsworth Longfellow*

Now you need to condense all the thoughts you've had about the report into a short objective - no longer than 3 sentences.

For example the objective for this book is:

To teach people, with various levels of experience, how to plan, write and present information as reports. To include a range of opportunities to practice and develop their skills.

Here's one from a report I had to write:

To give an overview of the current situation of personal development plan (PDP) implementation and to put it in the context of the national perspective and history of PDP at the University. To use this information to outline possible future directions in the developments of PDPs.

So your turn now:

> **Exercise:** For your report write an objective that covers everything that should go into the report.

# Agreeing the objective

*"You don't write because you want to say something,
you write because you've got something to say."*
*F.Scott Fitzgerald*

One important use of an objective is to return to whoever asked you to produce the report and check with them that this is their intended outcome.

This simple process of confirmation of the objective will save much more time later, as it will lead to the right information being included in the final report.

Where possible meet with the person involved or at least have an email conversation about the report.

Before you discuss the objective with them, or if you are unable to contact the person directly, you need to try imagining that you are the person receiving the report on your desk.

**Exercise: Pretend that your report is complete and then that you are the person who will receive your report.**

Imagine going into the room of the person who will receive the report with it in your hands, then swap positions with them and imagine their thoughts.

Think about why the report has arrived on my desk.
What else is on my desk to do today?

Do I need to make a decision based on the report?
Do I need to send it on to someone else?
How long is the report?
How long will it take me to read?

Am I able to talk to the person who has sent it?
What impression do I get on my first flick through?

How important is it to me?

**Notes:**

# Chapter 2: Planning and Resources for your Report

"Let our advance worrying become advance thinking and planning." Winston Churchill

# Chapter 2:
# Planning and Resources for your Report

Once you have got an agreed objective that makes it clear what the report should cover then next step should be done very quickly. This is to brainstorm the possible content. It is important to do this very early on, even if you know you won't be writing the report for a while, because you need to allow time to collate the information sources. This is particularly important when you might need to request information from other people as you don't want to be panicking at the last minute about a missing bit of data.

So the first step is to allow yourself to brain write (like brainstorming but done on your own) all the possible content of the report.

*"Knowledge can be enormously costly, and is often scattered in widely uneven fragments, too small to be individually usable in decision making. The communication and coordination of these scattered fragments of knowledge is one of the basic problems-perhaps the basic problem of any society." Thomas Sowell*

# Brain Writing

The idea of brain writing is that you allow your mind to free wheel and write down everything and anything that could go into the report without any censorship. So anything that occurs to you will be written down at this stage and then later you will decide which parts you need.

You are now into the creative part of organizing the information for other people and you need to be careful to use a method that allows you to think about what the information should be before organizing it. The mistake many people make is that they plan the content and structure together and this can lead to a lack of flow for the reader and can also cause a mind block and then prevarication on writing.

Following are two suggested methods for making sure that you think about the content and structure separately. Both of them have the same three steps:

1. Firstly, you dump your brain out onto paper.

2. Secondly, you evaluate the content and decide what should actually go in.

3. Thirdly, you decide how the information should be connected and structured in the report.

Now step 1 must be done on its own, steps 2 and 3 can either be done separately or they can be done together. The more complex your report the better it will be to do 3 separate steps.

## Method 1: File card frenzy

### Step 1: Brain writing

First, read your objectives to remind your mind of the purpose of the report.

Get sticky notes or pack of small file cards (if you don't have any cut printer paper into quarters).

Count the cards out into piles of 10. For short reports maybe 3 piles, for huge reports perhaps 8. Have spares in case you keep going with the ideas, if you use the spares again count them into 10s.

Now you are going to write one idea only onto each card for something that could go into the report. Do not censor your thoughts, anything and everything you think of should be written down. Remember you are going to evaluate the ideas later. The ideas can be a big one such as a section title or a small one such as a one little fact you need.

Take a pile of 10 and just start writing.

As you finish each pile of 10 pick up the next pile and again keep going. If you start a new set of 10 you have to keep going until all of them are filled.

The reason you've put them into piles of 10 is because this will help you empty your mind - if you have a fixed number of cards your brain has the task of filling all of them and will sweep out the corners of your brain.

## Step 2: Evaluation

The following day, read your notes on the answers to the 8 questions and your main objective again, then go through your total pile of thoughts and put them into 3 piles. If you can get colored paper to put each pile on then that will be helpful.

1. GREEN: 100 % certain this needs to be in the report.

2. AMBER: Not sure whether it should go in or not

3. RED: Definitely not needed

Keep piles 2 and 3 for now but just work on step 3, the structure, with the green 100% certain pile.

The reason why you do this process the following day is to give your brain incubation time. Fantastically, if you have a problem your brain carries on processing it for you during physical activity and sleep. Also, the separation in time will make you more objective about what is actually needed, remember avoid that groan factor, the minimum content for your purpose is what you're aiming for.

### Step 3: Grouping

Now there are 2 options for this part. You can either sort the ideas out on a table or, particularly if you used sticky notes, you can make a big map on a wall. Doing it on the wall is quite fun and gives you a different perspective. If you've used file cards then you'll need sticky tack.

List the main categories you think your information falls into and write each one onto a separate piece of printer sized paper. Have an extra one for uncategorized. Now sort your file cards out into the categories.

Alternatively, if you're not sure about your categories then place roughly the number you'll need as blank pieces of paper and then as you come to each idea on your file card ask yourself does it belong with any of the ideas already on the wall/table or does it need a new category. When you've finished then label each category.

When you have finished all your green 100% certain ideas then go through the pile of amber not sure ideas and see if you feel they now should be added to each category. Try not to add any unless you really feel they are going to contribute to fulfilling your objective. As a final check just go through the red pile once more and check you still agree that there is nothing there you should include.

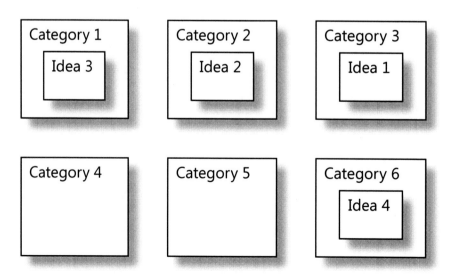

# Method 2: Sun diagram and mindmap

This is a great technique as it lets ideas really flow and allows you to see both the detail and overview of your work on a single sheet of paper.

### Step 1: Sun diagram

First slowly carefully read your objectives to remind yourself of the purpose of the report.

Get a large piece of paper, A3 or tabloid size is best. If you don't have any, then tape two pieces of printer paper together on the back.

You start with the main theme of the report and a little picture of what the report is about in the centre to get your creative ideas flowing. Circle the center and then draw simple lines radiating from the center just like a kid's picture of the sun.

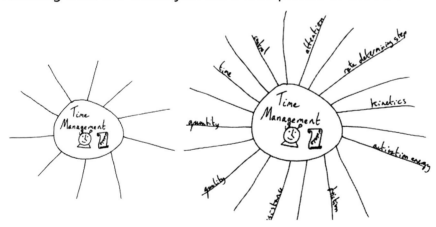

Fill in any ideas you associate with the topic, writing along the lines (if you write at the ends it will be very difficult to read it).

Then when you've used all the lines add another full set all the way round. Again fill all these lines and add another complete set in between them.

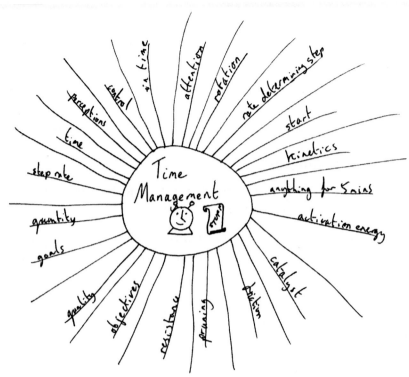

When you've finished it may not look pretty but it is amazing
how much this activity accesses the depths of the connections in
your mind to draw out all the information on a topic.

Now the reason you do it in stages is that you get a nice run of ideas that flow well, then as you go round again you are still flowing with ideas and they are now getting connected on paper with other things that may not have been natural connections. Thus, when you come to organize the data the next day you will see new and creative connections within the data that spark understanding and creativity.

You draw a complete set of lines each time because the blank line is like an unanswered question, your mind hunts for that little bit of extra information to complete the task of filling in every blank line.

## Step 2: Evaluation

The following day read your notes on the answers to the 8 questions and your main objective again.

The reason why you do this process the following day is to give your brain incubation time.

Look at all the things on your sun diagram, using a different color pen tick or highlight any that will definitely need to go into the report, using a different pen cross any that are definitely not needed.

## Step 3: Organizing the data with a mindmap

Now you're going to make a mindmap of the report. They are very easy to do and can help a great deal with planning. For examples of mindmaps please visit www.UoLearn.com and you can download a free booklet.

| | |
|---|---|
| Get a fresh piece of large (A3/tabloid) paper and turn it to landscape orientation. | A3/tabloid Landscape |
| It's nice to use colored pens to draw mindmaps as it helps to separate out the ideas. If you have them the best sort of pens are the multicolored packs of fineliners. | |
| Again start with your keywords and a picture in the centre. | |
| From your sun diagram choose your main sub-themes, allocate each one a color and draw a branch for each one and make sure they visually connect into the central picture (you should think of it like a tree the branches need to connect to the trunk). Now go round your sun diagram and add the ideas as sub branches to your main ones. Tick off the ideas from the sun diagram as you use them.<br><br>or<br><br>Look at your sun diagram and choose a category. Make a branch and start taking ideas off your sun diagram, crossing or ticking them off as you go along and write them onto the sub-branches.<br><br>When you've think you've finished one main idea then add a new branch for the next idea with a different color. | <br>or<br> |
| Add in as many sub-levels as you need, if you run out of paper use sticky tape to add more. | |

Example mindmap:

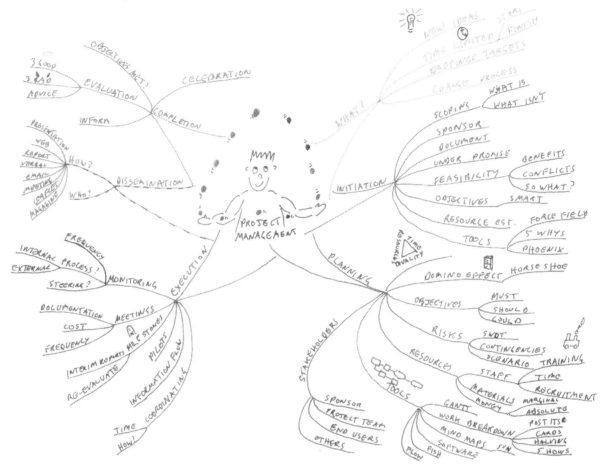

---

**Exercise: Brain writing your report content.**

Re-read your objective for the report you have chosen. Pick one of the two methods for brain writing (either file card frenzy or mindmapping). Give yourself a minimum of half an hour to empty all your thoughts about your report onto paper.

Remember, that you should allow free association and not censure any ideas in step 1. Keep going until you run out of ideas. The next day then work through steps 2 and 3.

# What resources do you need?

## Getting organized

Whichever brain writing technique you used, you should now have an overview of all the information that will go into your report. Now is the time to organize and collect all your information sources. Allocate a draw or folder for any physical data that you need for the report such as books, other reports and printed off notes. Make a folder on your computer for any electronic data, if necessary. If you are likely to be collecting a lot of the data via email correspondence then make yourself a folder within your email account just for this report.

Go through your file cards or your mindmap and find every single fact and piece of information you will need. Tick them off as you locate them. Do not delay this step, even if your report is not due for weeks and you are not going to write it for a while. There are two reasons for this, one is to give your brain an overview of the extent of the data and the other is to ensure that you are not rushing around and panicking right before your deadline. If you are dealing with a huge amount of data then it's a good idea to number each section (which will be a pile of file cards or a branch on your mindmap) and then have separate physical/electronic subfolders for each section and write on paper sources and rename electronic sources to include the section number.

If the data is usually held elsewhere either make a note in your folder of where it is stored (and check now it's there) or make a second copy. It is especially important at this stage to request any information you need from other sources

You should now have a list of the content for your report you now need to identify what information is missing.

**Exercise: Go through the cards or mindmap and identify any missing pieces of information.**

| What resources do I need? | Where/who can I get them from? |
| --- | --- |
|  |  |

## Gathering the Information

### From other people:

If you are asking for other people to provide information then be very explicit about your needs – tell them the following:

➢ Exactly what information you need.

➢ What format you'd prefer it in (for example if you're asking for data and they are likely to have it in an excel spreadsheet ask for that rather than a print out).

➢ Why you need the information.

➢ What date you need it by, which should be well in advance of when you need to write the report. It is best not to give people too long, take into account how much work is involved for them and then just give them a small amount of extra time. If you say you don't need something for two weeks people are likely to forget. If you say 2 days you're likely to get it.

Make a note in your diary to follow up whether the information has come in.

### From a library:

As soon as you identify that you need a particular paper or book then go and get it. Even if you don't have time to read it yet, make sure you've got it. Make a note in your diary of any due dates for the return of books.

### From the internet:

It is very easy to waste time on the internet. Have a concise list of the exact information you need and don't let yourself go on diversions. As you find it copy the data into a word file, along with the web address and the date that you accessed the information. You could make a special folder in your bookmarks and bookmark each site that is useful.

If you'd like to get super organized and have a huge amount of data why not use the data form below for your resources. It can also be downloaded as a word file from www.UoLearn.com

| Resource required | Date needed | Person (email)/ Web address/ Location | Date received | Where did I store it? (directory/ email/file draw) |
|---|---|---|---|---|
| | | | | |

## Reading the information:

[Extracted from the book Speed Reading Skills Training Course, www.UoLearn.com.]

Now if you've collected a huge amount of information you are faced with your own groan factor. The way to approach this is to use simple techniques for effective reading. The idea behind these techniques is that you set your mind onto hunt mode then you get an overview of the content and only then do you drill down into the detail.

1. Your mind needs to have objectives for reading.
   It helps to sort out what is important.

An objective should be in the form of a single sentence, possibly two, and needs to be in the format of a question. You should have no more than 3 objectives for each piece of reading. Here are some starting questions to help formulate the objectives.

➢ **Your motivation:**

Why am I reading this?

What am I going to use the information for?

➢ **Context with other people:**

Who else is going to read the material?

Who else will need the information but won't read it?

Why do they need the information?

➢ **Your personal context:**

What do I know already?

What do I need to know about the topic?

What don't I need to know?

The objectives need to be written as questions to get your mind into the hunt mode where it can't rest until it has found the answers. The six questions journalists use to tell a story are, who, what, when, where, why and how, your objectives should start with one of these.

2. Read the introduction – this will usually tell you what the material is for and what's in it. You are beginning to allow your brain to fit the reading material into context with the rest of your knowledge.

3. Read any summary.

4. Preview every page for about 2 seconds, this is almost like photographing each page. You look at the page, turn the page then look at the next one. As you are doing this you are both evaluating the reading material and starting to get an overview to fit detailed information in later.

5. Be very disciplined – keep going even over any interesting parts. If there is anything particularly important pop a sticky note on that page but keep moving on at this stage and return to it later.

6. Think about your objective and the following questions:

➢ What don't I need to read?
➢ Which part is most important?


7. Now revisit your objective and check it's still what you need from the material.

8. Read the sections relevant to your objective, using a pen under the text to make yourself read faster.

9. Have you fulfilled your objective?  If yes, then stop.
   If not, take a break and do something different (preferably over night).

10. Repeat the steps.


After you've read the material for the report, if you can, leave a couple of days before starting to write it. This is incubation time that allows your mind to make sense of the material whilst you get on with the rest of life.

**Notes:**

# Chapter 3:
# Organizing your Report

"First comes thought; then organization of that thought, into ideas and plans; then transformation of those plans into reality. The beginning, as you will observe, is in your imagination." Napoleon Hill

# Chapter 3:
# Organizing your Report

## Getting the report into the right order

You should have now reached a stage where you've got all your information and have an idea of what sections you might include, so now it's time to decide what order to put everything in and to list all your section headings.

Reports can be put into three main categories:

➢ Information reports – collated information

➢ Research reports – you've collected data and someone (probably yourself) has analyzed it

➢ Proposal reports – information is presented with proposals being given so that a decision can be made by the reader.

Below are the full lists of the types of sections for each of these styles. You may not need all of them in your report.

| Information | Research | Proposal |
|---|---|---|
| Title page | Title page | Title page |
| Executive summary | Executive summary/ Abstract | Executive summary |
| Contents and tables of figures/diagrams | Contents and tables of figures/diagrams | Contents and tables of figures/diagrams |
| Aim | Aim | Aim |
| **Introduction** | **Introduction** | **Position** |
| **Categories of information** | **Method** | **Problem** |
| | **Results** | **Possibilities** |
| **Summary** | **Conclusion** | **Proposal** |
| Index | Index | Index |
| Bibliography | Bibliography | Bibliography |

You can see there are common factors in all three of the report styles. We are going to look at each one of these in more detail. Where elements are common to the reports we'll cover them just the first time.

The navigation aspects, such as the contents, index and lists of figures will be covered in chapter 5 as they will be one of the final things that you sort out.

# Information reports

The vast majority of reports fall into this category. You, the writer, have collated information sources and are going to share them with other people.

**The sections an information report is likely to need are:**

➢ Title page

➢ Executive summary (abstract)

➢ Aim

➢ Introduction/overview

➢ Categories of information

➢ Summary/recommendations

➢ Bibliography of information sources

**Title page**

It is really important that you have a separate title page. It shows the reader that the report is important. It is the space to make sure that important identification data is present. Even if the rest of your report is to be printed in black and white it is a nice idea to get your title page done in color.

A title page should have:

➢ The title

➢ The date of publication

➢ Either your own personal contact details or the departments, depending on the type of report.

➢ A picture, suitable for the report, you can get these very cheaply to use in any media from sites such as www.fotolia.com, www.istockphoto.com or www.shutterstock.com.

Exercise: Design your cover sheet.
For your report write your title, find a picture and decide what contact details should go on the front page.

**Executive summary**

Now, remember the catch phrase for writing reports,

> "It's all about the reader."

Your job as the writer is to help your reader extract the information they need as quickly and easily as possible. So a summary right at the start of the document is a vital component of this. It helps the reader get an overview of the content before they put the detail in and also helps them identify which parts of the document will be useful to them. As a guideline I'd suggest any document over 5 pages long should have a summary at the start. Depending on your audience, label your summary as an executive summary or abstract.

It is very important that your summary contains only the same information as the report, don't add anything to it that isn't in the main report. Executive summaries should only be written after the report is completed. The original purpose of them was to be removed from the main report and circulated separately to the management team to help them give an overview of what was going on. If you think your summary is likely to be circulated separately make sure there are title, contact details and information of how to obtain the full report.

The 2 questions that will really help you to decide the content are:

➤ What's the minimum people need to know?

➤ What would I tell them if I only had 2 minutes to talk about my report?

Exercise: For your report think about the 2 questions above.

**Guidelines for executive summaries:**

✓ Reports over 5 pages should have an executive summary, depending on your audience you may wish to call it an abstract.

✓ A suitable length is a maximum of 5% of the length of the report (excluding appendices) and as a rough guide for most reports it is best to be able to get it on one or two pages.

✓ It must summarize the report and be able to be read and, if necessary, circulated separately.

✓ They are usually organized in the same order as the report itself.

✓ They are written after or alongside the main report.

✓ Only information given in the main report should be included.

✓ The emphasis can change slightly, for instance a 10 page section of your report may only be 1 sentence in your summary. A 2 page part may be 3 sentences.

✓ Look at the titles of your subsections and beginning and final sentences of paragraphs to start outlining your summary.

✓ Find key words and use them to organize the draft of your summary.

Exercise: Look at the executive summary at the start of this book. Compare it to the actual contents. Visit www. UoLearn.com and find the report writing section where there are some examples of reports with executive summaries for you to look at.

## Aim

The aim is usually just your objective, perhaps rewritten slightly, for the report. It should tell your reader what you report will cover and be very precise so they can quickly identify what is not likely to be in the report.

## Introduction

This is an optional section and is only necessary if you think your readers need one of two things:

➢ Help understanding the context of the information.

➢ Help understanding the structure of your report.

Introductions should be very short and definitely not the entire history of the report and your journey through writing it. Remember before you write an introduction re-read your objectives, if it doesn't help meet them don't put it in. If you do include an introduction don't just call it introduction use a helpful and informative title to help people understand the purpose of the section.

Exercise: Think about the last few reports you read. Were the introductions helpful, if so why and if not did they need them at all?

### Categories of information

This is the biggest and most important part of an information report. Hopefully you've already done most of the work for this section, in the brain writing part of the book. The categories are probably the branches on your mindmap or the titles of your groups of file cards. Some possible types of categories could be:

➢ Date order
➢ By department
➢ By group of people (eg. course at a college, type of job roles)
➢ In order of importance to you reader
➢ By size
➢ By geographical location
➢ Alphabetical (not a usual choice as it doesn't often take the reader into account)
➢ The order to undertake tasks (such as you might get in a self help manual or report that teaches a skill)
➢ In an existing order that your reader will understand (such as catalogue order, year order at a college, same order as last 10 reports etc.)

In each case you need to decide the section order that will make it easiest for your reader to follow.

Exercise: Take your category headings from your brain writing and organize them into a logical sequence for your reader. Once you've done that think about your reader opening up the report, will they find this order easy to follow?

One of the most important parts about the categories is that you make the titles long enough to tell your reader what is in that section but also what is not in that section. Section titles should be specific enough that the content couldn't overlap with any other section in the report. Usually they have plural nouns.

For example:

Rooms that need refurbishing in building XYZ.

Regulations for the exam board for the chemistry degree.

The reasons for considering department A for the first stage of the project.

If your sections are going to grow to many pages each then you should also put in subsections, again the same ideas apply to sub-section titles, make them descriptive not just single words.

The whole point of structuring your report into sections is to help make sign posts for your reader to help them decide what to read and in what order.

Exercise:
What headings are you going to give your categories?
Do they tell the reader both what to expect in that section and also what won't be there?
Is the order a logical one for your readership?

### Summary/Conclusion

The summary should be a short part of the report that pulls together the information from the categories. It should be used to emphasize the parts that you feel are most important for the reader to remember and help to give an overview of how the reader can now use the new information that they have.

### Bibliography

Again this is an optional section, you may not need one at all or it may be extra material that could be available on the web. You could name it something more appropriate to your readership such as further reading, references, resources, links or book list. You need to decide on a layout for each type of resource and make sure you use it. Depending on your readership this may want to be very formal, such as the Harvard system (see www.libweb.anglia.ac.uk/referencing/files/Harvard_referencing.pdf for a comprehensive guide) or you may wish to add commentary to each resource telling the reader a bit more about it. Whichever technique you use, you need to make sure that you give enough information for your reader to be able to locate the reference.

### Book references:

Example from the Harvard system for a book, author, date published, title, location of publishers, publishers name.

Baker, H., 2010. *Speed Writing Skills Training Course*, Lancashire, Universe of Learning

Then if you referred to the book in the text you'd put (Baker, 2010).

A less formal method would be:

**Successful minute taking - meeting the challenge,**
Heather Baker, ISBN 9781949370387

## Journal references

These would definitely use a formal system as any reader looking up a journal reference would be used to the Harvard system and expect to see it.

author names, date, article title, journal, volume number, (part number), page number.
This time you'd put the journal name in italics not the article title.

Greenhall, M.H., Lukes, P.J., Petty, M.C. Yarwood, J. and Lvov ,Y. , 1994. The formation and characterization of Langmuir-Blodgett films of dipalmitoylphosphatidic acid. *Thin Solid Films* 243 (1-2), pp.596-601

(Yes this really was me, in my previous life as a chemistry lecturer!)

## Web references:

In the Harvard system you'd put author, date, page title, web address, date accessed.

Baker, H, (2009), *10 uses for speed writing*, http://www.uolearn. com/speedwriting/10usesofspeedwriting.html, Date accessed 30/06/10

Less formally you might just give the title and web address.

Whatever you choose make sure you're consistent in both information and the style of the text.

# Research reports

A research report not only presents data but it also analyses it. The data is often very new and recently collected either by yourself or a colleague.

Examples would be:

A survey of needs in your department.

An analysis of sales trends over the last 3 years.

A report into the testing of a new price of equipment.

**The sections a research report is likely to need are:**

➢ Title page

➢ Abstract (Executive summary)

➢ Aim

➢ Introduction

➢ Method

➢ Results

➢ Conclusion/Discussion

➢ Bibliography of information sources

We're just going to look at the additional sections to the information report. If your audience has an academic background you may wish to re-label the executive summary as an abstract.

## Method

This is the section where you describe to your reader how the data was collected and analyzed. It is to help give credibility to the information you are going to present later so your reader can decide whether to believe the data or not. It can also be used to help someone else reproduce your research.

For a research paper to a formal journal you'd always include a method section. However, unless you think there will be a problem with either your reader understanding the data or them believing it, a method section should be put in as an appendix or addition web based material. This is because they are often lengthy and usually the reader doesn't want to know how you achieved the results, they just want the answers.

Again as with the categories, if you do include a method section don't just call it method. Use a nice descriptive tile to help your reader understand what the section is about so they can decide whether they need to read it or not.

Examples:

Techniques used to design and distribute a questionnaire on departmental needs.

How the data was collected and analyzed from the sales department.

The method and equipment used to test the thingymabob.

**Results**

This is where you give the results of your analysis. This section would be included in virtually all research reports. Again your title should be descriptive.

Examples:

Departmental requirements for the next 6 months.

Predictions of future sales growth based on the last three year's data.

The strength and longevity of thingymabobs.

It is nice if you can present results graphically where possible, see www.UoLearn.com for some ideas on presenting numbers visually.

Remember you don't need to give every single little detail of what you found. You should select only those relevant to your objective. If you want the full data to be available, again upload it to the internet or your internal network. This section should only be the outcomes of your analysis, the analytical process itself should be in the method.

**Conclusion/Discussion**

This is where you draw together the outcomes of your research and suggest ways that it could be used. It should connect back to the original aim of the report.

# Proposal reports

A proposal report presents a problem that needs solving, explores potential answers and then puts forward possible solutions. They are often the trickiest to write as you are helping a decision maker choose a solution that may have lots of implications for the future of your department/company.

Examples:

A report on a safety incident which gives solutions to avoid further reoccurrences.

A proposal for which site to choose for a new factory.

Which one of three new computer software systems should your company implement.

**The sections a proposal report is likely to need are:**

➢ Title page

➢ Executive summary

➢ Aim

➢ Position

➢ Problem

➢ Possibilities

➢ Proposal

➢ Bibliography of information sources

Again as for the other types of report your actual titles for the sections will be far more descriptive.

The 4 ps are a familiar pattern from stories we all know.
Position: 3 pigs build different types of houses.
Problem: A hungry wolf.
Possibilities: Which house to huff and puff first?
Proposal:  Go for the least solid structure for an easy meal.

59

### Position

This is where you are now and the history behind the problem. Depending on your readership you may be able to combine this with the problem section.

### Problem

This is where you have to outline why change is needed - what is wrong with where you are now and what benefits a solution could bring.

### Possibilities

This is often the biggest part - you give an impartial account of the range of possibilities that are being considered without favoring or overly criticizing any particular one.

### Proposal

This is where a proposal for a way forward to solving the problem is made.

Now you have to be very careful about your role as the report writer. Are you involved in the decision process as a key player? If so it may be OK for you to decide on which possible solution to implement. However, if you are only putting together a report for other people to make a decision then you need to be very careful not to put your own opinion into the way you weight and discuss the proposed answers to the problem.

Now, whilst these four sections are usually in a proposal report you might like to think carefully about what order they should be in. The traditional way is as outlined above where you are taking your reader through the same thought process as you (or your colleagues) went through to get to the answer. I call this the Miss Marple approach. [In Miss Marple you follow her progress in solving a crime alongside her and nobody knows for sure who did the deed until the final scene.] It's very useful if you think it will help your reader decide on the options by taking them through the full problem solving process.

However, if when you look at your objectives you realize that your reader would not need to follow your progress through the problem solving process, then you might want to consider the fast track approach where you give them the proposal for options first and then back them up with the reasoning afterwards. I call this Columbo technique, where you show whodunnit and then you prove it. [In Columbo you see who did the crime and how and then follow Columbo as he builds the case against the criminal.]

Reports that present the proposal towards the start tend to be more direct and often shorter. Often by presenting the answer first the position, where you are now, becomes self evident and doesn't need its own section. When you present all of the options and then choose the best you tend to explain each in far more depth covering how you would implement each one and the pros and cons. By presenting the best proposals first you automatically spend less time explaining the ones that aren't going forward.

**Traditional Miss Marple approach**

**Position**
Where we are now?

↓

**Problem**
What's the issue?

↓

**Possibilities**
What are the options and how do we
implement them?

↓

**Proposal**
Which solution/s are best?

An example proposal outline for office reorganization would be:

**Position:** We have offices in 5 cities, each has separate accounting offices.

**Problem:** Communication between the offices is poor and mistakes and omissions are being made, it is also costing money to repeat skills.

**Possibilities:** Relocate all the accounting departments to city X.
Relocate to 2 locations X and Y.
Use technology to link all the depts together, including a bulletin board to keep everyone up to date and web conferencing.
Do nothing.

**Proposal:** Due to staff problems of relocation the technology route is probably best, therefore this report proposes an in-depth study into the best technological solution.

**Columbo - reverse plan**

### Proposal
Which solution/s are best and
how do we implement them

### Problem
Why they solve the problem

### Possibilities
What were other options

### Position
By the time you're this deep
into the report your starting
point will already be clear

An example proposal outline would be:

**Proposal:** It is recommended that we invest some money into technology to connect our accounting offices. This way we can avoid repeating skill sets.

**Problem:** This would solve the current problem of omissions and minimize mistakes.

**Possibilities:** We also considered the options of relocating the personnel but decided that it would be too much of a disruption to personnel.

**Position:** Already evident to the reader from the rest of the report, so not needed.

**Exercise: Consider your report now.**

Which type of report is it?

........................................................................................................................................................

Using your brain writing information, what are the sections you need in it? Based on your thoughts about your readers, what order should the sections be in? Are there any that could be moved to appendices or made available separately to the main report, as for instance back up web based material? What descriptive titles are you going to give your sections?

| Section type | Order | Descriptive title |
|---|---|---|
| | | |
| | | |
| | | |
| | | |
| | | |
| | | |
| | | |
| | | |
| | | |

# Chapter 4:
# Presentation of the Report

"The pages are still blank, but there is a miraculous feeling of the words being there, written in invisible ink and clamoring to become visible." Vladimir Nabakov

# Chapter 4: Presentation of the Report

## Making your report memorable

If you've been following along with the exercises you should now have a plan of your report. It's now a case of getting your writing done. Everyone has their own writing style but there are some simple principles that can make your work not only easier to read but also more memorable by how you design and write the report.

There is a lot of research on memory and some of it can be applied quite easily to writing.

## Positioning of information:

People tend to remember:

✓ Starts

✓ Ends

✓ Anything unusual

Now if your report is long then it is particularly important to think about the fact that your reader may not remember much from the middle of the report. One way of approaching this is to have very clear section headings so the reader gets lots of mini starts and ends throughout the report. Another method is to put something unusual in such as a change of font, an illustration or a funny quote. Of course many people use this principle in reverse, bury the material in a repetitive style in the middle of a report with no visual stimulation and you can help them to forget it!

## Amount of memory

✓ Working memory is 30 seconds long

✓ There is only space for 7 items

People have a short term memory capacity called working memory that is very short (30 sec) and is the space you have to juggle ideas and integrate them with the rest of the information you already know. Now in terms of writing, the interesting thing is that people can only hold 7 items in their working memory at any one time. So you need to make sure that you give them no more than 4 things to think about, preferably 3, so that they still have space in their working memory to make connections.

Now, when writing in short paragraphs most writers tend to automatically keep the number of ideas down to 4. However, often in reports there will be bulleted lists. If they are arranged so that the reader already has to engage with them one at a time (such as a sequence of actions - see the previous section on speed reading) or as a checklist (see the section on proofreading) then it may be OK to have one long bulleted section. However, for factual information that requires no action by the reader you need to be careful how you layout the list or your reader with either not take it in or may be selective about the items that they remember.

Limit your ideas to a maximum of 4 at once, you don't necessarily need to have sub titles, just laying the list out with spaces after every 3 to 4 items is enough to help the reader think about those then move onto the next group. If it is possible try to make the group of 3 make sense as a group but even if they don't just having a space will help your reader understand your points better. Have a look at the following examples to see which one helps you to take in the information better.

Example layouts:

**Time management tools:**

- ➢ To-do lists, keep them up to date
- ➢ Diary, paper or electronic
- ➢ Gantt charts for your projects
- ➢ Tidy up, allocate 5 mins per day
- ➢ Prioritize, divide your tasks up into A, B or C
- ➢ Rotating tasks, make yourself change tasks
- ➢ Control your email,  only check your email 3 times a day
- ➢ Go to another room, get away from chatter
- ➢ Divert your phone or let the answer machine take it
- ➢ Say no to new ventures unless you've just finished one
- ➢ Delegate, what do you do that someone could do for you?

**Time management tools:**

- ➢ To-do lists, keep them up to date
- ➢ Diary, paper or electronic
- ➢ Gantt charts for your projects

- ➢ Tidy up, allocate 5 mins per day
- ➢ Prioritize, divide your tasks up into A, B or C
- ➢ Rotating tasks, make yourself change tasks

- ➢ Control your email, only check your email 3 times a day
- ➢ Go to another room, get away from chatter
- ➢ Divert your phone or let the answer machine take it

- ➢ Say no to new ventures unless you've just finished one
- ➢ Delegate, what do you do that someone could do for you?

## Attention

If your report is very long you need to consider the fact that people's attention span is quite short. Most people's attention span, for material that is not interesting, is 20 mins. After only 15 mins people start to lose attention and after 30 mins most people enter the snooze zone and will automatically stop and take a break as they start to drift off. If someone is not paying attention they will not be making connections with their memories and will lose understanding and will not remember what they've read.

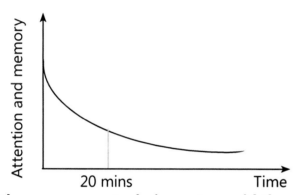

**Now there are some techniques to avoid the snooze zone:**

➢ Keep the report short enough to read in 15 mins!

➢ Break the report up into very clear blocks, perhaps separated by visual indicators such as pictures and title pages (like the chapter starts in this book). This means that there is an easy to identify place to have a break.

➢ If appropriate add humorous parts in (obviously this is usually not an option but sometimes you can do it).

➢ Change them into a different mode of reading - the hunt mode by asking question within the text. These could be your heading titles. According to advertising marketing research using a question as your heading can get a massive change in how many people then go onto read the main text.

Why do people read more?

Because our brains hate unanswered questions!

> Exercise: Go back and look again at your section order and titles, could you use any of the information on memory to improve any of them? Are they in the right order?

# Layout style

Although people think mostly about the content of their report, it is important to understand that making it easy to read is almost as important. Again, it comes back to avoiding the groan factor.

If you look at the next page you'll see two examples of the same text, believe it or not both are in 12 point font. The first is in Times New Roman and the second in Arial. Your first concern is your reader, if you make it hard to read you'll immediately get a negative attitude to the report, no matter what information you put in it.

Make sure that you plan to leave pages for any navigation components such as contents, index, figure lists, table list etc. This is to make sure your page numbering will be correct. These aspects are very important, if you think of your report as a website they are the navigation buttons that help people jump to exactly the information they need.

Exercise: Look again at the reports of other people, there are some links on our website and you should have gathered some of your own. What is the best layout you can find. Could it be improved at all?

Which one would you be more likely to read?

**Closed lists – do it tomorrow**

One of the biggest problems people face when they try to manage their time is what gets done when. Many people use to-do lists where a mixture of tasks is thrown onto a piece of paper. The trouble is that they keep growing and there is no absolute time constraint on when they are completed. A closed list can not be added to once you've written it. Examples of a closed list would be 10 reasons why global warming is taking place, 5 things I will get done today. So the last 5 to 10 minutes of each working day or revision period should be used to write a list of what you'll do tomorrow. This list should only include what you realistically think you can get done and should include a reasonable amount of time for unscheduled interruptions. Once this list is written draw a line under it or box it in. Nothing can now be added to this list. You will have succeeded if you completed everything on the list

## Closed lists – do it tomorrow

One of the biggest problems people face when they try to manage their time is what gets done when. Many people use to-do lists where a mixture of tasks is thrown onto a piece of paper.

The trouble is that they keep growing and there is no absolute time constraint on when they are completed.

A closed list can not be added to once you've written it.

### Examples of a closed list would be

➢  10 reasons why global warming is taking place,
➢  5 things I will get done today.

So the last 5 to 10 minutes of each working day or revision period should be used to write a list of what you'll do tomorrow.

This list should only include what you realistically think you can get done and should include a reasonable amount of time for unscheduled interruptions.

Once this list is written draw a line under it or box it in. Nothing can now be added to this list. You will have succeeded if you completed everything on the list.

## Layout checklist

➢ **Is there a feeling of space?**

White space on the page can be the most important aspect of your layout. Hopefully, you've found this report writing book easy to read. A lot of that is down to limiting the amount on a page. Not only does it look better it also helps with understanding (see the memory section above). Often people try to save on copying by squeezing their report into fewer pages. It's not worth it, remember the motto - it's all about the reader, would they prefer 1 less page or an easy to read report?

➢ **Is your font straight and at least 12 point?**

The dyslexia association recommends that you use a sans serif font (non curly one) that is at least 12 point. I think that if it makes it easier for people with reading difficulties it must help everyone else.

Examples (all 12 point) are:

Calibri looks like this (the new Microsoft default font).

Segoe is what this book is written in.

Arial is the most well known of the straight fonts.

Comic sans is a friendly font (used in the exercises).

Verdana is another straight font.

➢ **Have you inserted page numbers?**

Word processing software allows you to add not only simple page numbers but things like page 1 of 5, date or the filename. These can be found under the Insert/Auto text or the Insert tab.

➤ **Do you have a consistent style throughout?**

If you use word then it has a very useful function to help you keep your styles consistent. In the older versions of word it is found under the format tab and is called styles and formatting. In the newer versions it can be found in the home tab towards the right. These allow you to click on a style and apply it immediately to a paragraph. It comes with default styles such as heading 1 but you can both alter these (use the drop down menu) or add new ones of your own. If you haven't used styles before the best place to get help is just put "styles in word" into youtube. Then watch and learn. I would highly advise using styles for your headings. There are a few useful functions when it comes to checking your work and making a contents page that rely on using styles.

Modern title styles tend to be larger fonts than the main text and bold. Underlined titles are rarely used now.

➤ **Do you need to number your sections?**

If your report is going to be long, or if people are likely to need to refer to sections a lot, then you should number your sections. You can use the automatic numbering system in your word processor, this is helpful in that if you add a new section it re-numbers everything but can be frustrating sometimes when it doesn't do exactly what you want. An example of a typical pattern of numbering would be as below but some people do mix letter and numbers for different levels. The important point is to ensure that in each section the same level has the same numbering system as the rest of the report.

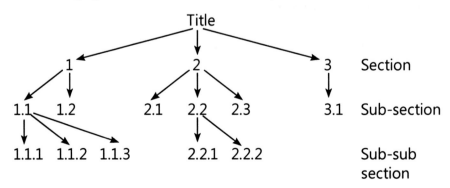

➤ **Are all of your section titles grammatically consistent?**

Now this is where using the styles in word really comes into its own. If you change the view to outline in the main view menu then it condenses the document down to the headings with + to indicate that you can expand to see what is underneath the heading. It's a bit like the folder view when you're looking at files on your computer. This is great for checking your numbering and that the titles match each other in their tense.

**Inconsistent titles:**

The analysis of the data from plant Y

Analyzing the data from plant X

Basically, if your verb endings don't match ing/ed etc it is possible you may need to look at your titles again.

➤ **What bullet styles will you use?**

It is best to decide on just a few simple bullet styles, you don't need to stick with the simple dots. Wingdings in particular has loads of nice characters. If you choose format/bullets/customize then you can choose anything you like. However, the proviso on this is keep it simple, choose perhaps one or two styles for your whole report. Don't choose a new bullet style for every list as it will make it look chaotic. In this book there are just a few.

✓ Tick for things that are definitely right or you want your reader to agree with.

✗ Cross for things that are definitely wrong (like the reverse brainstorm).

➤ Arrow for general points and questions.

Some others from Wingdigs to consider are:

Exercise: Spend 10 minutes investigating the capability of your software for having consistent styles for your headings. You can download our example file from the report writing resources section of www.UoLearn.com if you use word. Try out the outline view and find out how to change the format of the headings.

Exercise: Now design yourself a template file with all the formatting for your headings and bullets built in. You can either save this as a template or just as an ordinary file. Remember if you save if as an ordinary file keep 2 copies as you might forget to rename it one day.

## Looking after your files

Now you're entering the main writing phase of the report you need to start to be careful with your file management. Each day you work on the file save it as the name with the date in the title. For instance this file is called report0929, tomorrow it will be called report0930 and when I finish the last file will be called reportfin. If you're doing a lot of work on the report on one day you might even want a new file every hour or so, I add v1,v2 etc to the file (for version) as I go through the day. So yesterday was a good day for writing and the file ended up being report0928v4. It is also a good idea to back your files up periodically onto a separate system, disc or memory stick. Every so often files do get corrupted and by saving it as a separate file each day rather than over writing the previous one means you won't lose so much.

# Procrastination

Are you avoiding writing?

Have you found yourself cleaning out the bottom draw of your filing cabinet?

Avoiding a task will increase the size of your activation energy to start it. If you find you're doing lots of make work things like dusting your plants then you're avoiding something bigger.

The way to tackle a big task is to break it down into lots of manageable smaller tasks. You should have a list of section titles for your report from the earlier exercises. Don't think about writing the whole report, just make it your goal to fill one of the sections. It doesn't matter which one, you've got to eventually write the whole report anyway so start with the one you'll find easiest to write.

You can also trick yourself by saying all I'm going to do is just open the file. Then all I'm going to do is write one sentence. Make it something you couldn't possibly fail to complete.

If you're still finding it hard to start then make yourself write for 5 mins only and no more. How hard could it be to just tap away at your keyboard for 5 minutes? Even if it's no good and you delete it later at least you've started on your journey.

Once you start you'll find you want to continue.

*"You can do anything for 5 mins!" Mark Forster*

# Writing good English

I am going to follow my own advice on this section - it's not something all of my readers will need so if you feel your general use of English needs brushing up then visit the report writing section of www.UoLearn.com where you'll be able to download a guide with useful grammar tips. You can also find information on types of diagrams to include in your report.

## Writing style tips:

> **Sentence and paragraph length**

Keep both fairly short. As a guide, you should have about 20-25 words per sentence and 4 to 6 sentences per paragraph. Any longer and you'll loose the reader's attention.

> **Acronyms and abbreviations:**

Give an acronym or abbreviation in full with the short version in brackets the first time it is used it the report. After that just use the short version.

World Health Organization (WHO)

If there are lots of these in your report have a separate page listing them all. Most of the time your readers will be familiar with them. However, have you ever read a report or book and been annoyed that the author assumes you know what the abbreviations stand for? Consider perhaps that your report may be read by a new starter, perhaps even someone who takes over your job when you get a promotion for writing a fantastic report. A good resource is www.acronymfinder.com.

> **Jargon and long words:**

I floccinaucinilipilificate long words, yes it's a real word, it means estimate as worthless! Keep it business like but clear and understandable. Filler words to make you seem learned and great, things like forthwith, hereof, wherein, henceforth are superfluous (as is that, use not necessary, it's much easier to understand).

> **Eliding**

Most of the time it is quite acceptable to use shortened versions of words such as can't, don't etc.

> **Gender related language**

Usually, unless you are writing about a particular individual the modern approach to gender is to use the neutral versions of words such as they, their, person etc. Sometimes people use both genders together he/she but it actually looks odd visually. Another approach is to alternate between he and she, but this is hard to keep track of. Whatever you choose, keep it consistent throughout your report.

> **Ambiguous words**

There are words in English that can alter their meaning by the inflection when spoken. Think about the phrase:

"I was a little concerned with your results."

Depending on the tone you could be about to get some friendly advice or get fired.

So when you are writing it is usually better to use a more precise word so that your meaning can't be mistaken. Some examples are replace:

Alter with improve or worsen
Affects with reduces or increases
Expressed an opinion with liked or disliked.

**Notes:**

# Chapter 5:
# The Finishing Touches

"Hard writing makes easy reading. Easy writing
makes hard reading." William Zinsser

# Chapter 5:
# The Finishing Touches

So once you've written your main report it is still not finished. There are a few more stages to go through:

✓ Editing

✓ Writing the executive summary

✓ Putting in a contents and index page if needed

✓ Proofreading

✓ Printing

✓ Distribute

✓ Celebrate (or collapse)

# Editing the report

Often people make the mistake of doing the proofreading and editing together. If you want an excellent report that is concise and easy to read then you must separate out the two phases.

Proofreading is just about looking for inconsistencies in style, spelling mistakes and typos. This could be done by anybody with good English skills.

Editing is very different it is about checking that the content itself is correct. It is revisiting everything you've written and comparing all of it to two standards:

➢ **The original objective**
You need to re-read the answers to the 8 questions at the start and the original objective then think about the following:

➢ Have you missed anything out?

➢ Is there anything extra in the main report that could either be cut or moved to an appendix?

➢ Is it in the right order for your reader?

➢ Does it fulfill the objective?

➢ **The look and length**
Obviously you don't want to make everything too simple, you're unlikely to be writing a report for young children to read but the easier on the eye and the shorter the better.

➢ Does it look right?

➢ Are the any places where you waffle or repeat yourself?

➢ Could you replace two sentences with one?

➢ Do you use too much jargon or long winded phrases?

Editing is vicious you need to think of it like taking a pair of giant scissors to your report and cutting off the excess baggage.

If you visit www.plainenglish.co.uk they have some great examples of condensing text and a 630 word sentence.

# Writing the executive summary or abstract

If you return to the information report section there are guidelines for writing the executive summary.

Remember the two questions were:

➢ What's the minimum that people need to know?

➢ What would I tell them if I only had 2 minutes to talk about my report?

Reading a report is a bit like doing a jigsaw puzzle. Many people read without setting questions (see the speed reading section) and just start at the beginning and go all the way to the end. This is like tipping the puzzle on the floor and just grabbing the first piece from the top and then the next and hoping to make sense of it. The trouble is you've got nothing to attach it to. Your role as a writer is to help the reader develop a strategy for reading. So most people, if they were doing a puzzle, would find the outsides, then find the easy to identify pieces like a red coat and then finally go for the very detailed pieces of sky. This is what the executive summary does, it gives the reader an overview of what is to come and makes a framework in their mind to fit the rest of the information into.

The other great thing about executive summaries, particularly if you are in an organization where they are not usually used, is that you'll get comments on how easy your report was to read and how useful the summary was. Try it and see!

This should be your final piece of writing, phew.

Make sure that you edit it too.

# Contents and index

If your report is long or complex you'll need to help your reader to navigate it with contents and possibly even an index. You need to think of how people are now used to navigating websites and give them different options for finding the relevant material.

Now if you've used the styles in word the content page is easy to generate. Go to Insert/Reference/Index and Tables or References/Table of Contents and follow the instructions. For the contents you can even have live links as well as page numbers, if the report is to be distributed electronically. The heading styles can be added or removed from your contents to give you the required detail for your readers.

Word will automatically take into account the re-pagination that will take place, when you add a contents page. However, other software such as Indesign may not.

If you are adding lots of navigation elements such as lists of figures, tables, acronyms etc, then you either need to leave space for each one or once inserted refresh each of them to reflect the new page numbers. Again if you've been careful to allocate each of these headings its own style, generating the tables is easy using insert, reference.

Adding an index is harder, you first of all have to go through your report marking all the words that you want indexed and then generate the index. It's quite a lot of work so be sure it is necessary before you do it.

**Other things you could also include to help navigation are:**

- ✓ List of acronyms
- ✓ Definitions of technical terms
- ✓ List of figures
- ✓ List of tables

# Proofreading

This is the final stage before sending the report out to the world. Please make sure you remember to allocate time to do it. There is no point in writing the world's best report and having five people email you with a daft typo.

It is best if you can make sure that you have time between writing and proofreading to allow your brain to change modes and to see your own mistakes. If you're up against a deadline then try separating yourself in space by leaving your normal workroom and proofreading elsewhere.

The best advice of all is to get someone else to proofread your report and do something else in return for them.

At this stage you should not be adding or deleting chunks of texts - you'll only add mistakes in. (I once did that and ended up with presentation kills not skills, oops!)

**Ideas for proofreading:**

✓ Get someone else to do it (best choice)

✓ Read aloud

✓ Print it out

✓ Do several proofreads, each time look for different errors, one for full stops, one for spelling, one for titles etc.

**You are checking:**

✓ Spelling

✓ Grammar

✓ Formatting errors (style changes)

✓ Punctuation

**Checklist for proofing:**

- ✓ Starting point - have you used the spell checker!
- ✓ Style consistency throughout
- ✓ Layout changes in positioning things on the page, like margins and pictures
- ✓ Spelling main text
- ✓ Spelling titles (often missed - do them separately)
- ✓ Full stops followed by capital letters
- ✓ Apostrophes
- ✓ Repeated or missing words
- ✓ Your own common mistakes (my worst is form/from)
- ✓ Titles on diagrams and figures
- ✓ Numbers on diagrams/figures/sections

See www.writing.umn.edu/tww/grammar/proofreading.pdf for a more extensive list.

Exercise: What are your weaknesses in terms of errors, are there mistakes you make frequently?
If so make sure you do a separate check for them.

Exercise: The following almost passed a spell checker, it only picks up two errors but there are loads more. See how many mistakes you can find (answers in the downloadable workbook on www.UoLearn.com).

**Committee Sutures**

Committees are the back bone of any large organization. They provide structure for many of our civilities in practice many a long hour is used at you're usual committee meeting debating this way and that. However its was my privilege to attend a meeting with a difference.

I went along as a replacement for a colleague to a teaching and learning committee meeting Never before have I been surprised by the structure of a committee meeting. There all the same you have an agenda you work though it there are the odd interludes of excited exchange's if you're lucky but never does anyone question the fact that you all sit around one long table.

This particular meeting a very senior member of the university was chairing it. Mostly it was as every other committee meeting... However half way through the chair gave each group a different section of the 'fascinating new teaching and and learning strategy He then asking us to split into groups of three and use a flip chart to list any relevant points on implications of the strategy to the university.

I was enthralled – here in a top level committee meeting someone had dared to practice what they preached and use well accepted teaching methods to help structure the meeting. Wow it might seem like a small change but It changed the nature if the interactions around the table instead of the same 4 or five people talking and making althea decisions we were all given c a chance to look in depth and make positive contributions to a real document that was going to effect the future of the university I really enjoyed the meeting and got to no a couple of the committee really well whereas usually unless you have other meetings outside the committee you get to know peoples names and jobs but not much else

So my advice to you is to think of other ways of structuring the communication in a committee – you've got a real chance to make help people contribute in different ways and as you can see they might even remember what happened to. It's not everyday that you feel your voice has really been herd at such a high level within a huge organization.

© Margate Greenhall

# Printing and finishing

Before publishing just do one last check through. Look at it again with fresh eyes, if it landed on your desk would a quick flick through impress you?

Make sure you know how long it will take to print your report. If it is going to a print room or external printers they could take quite a while, possibly 4 or 5 days. If you are just running them off your printer or photocopier make sure you do it at least the day before the deadline - the gremlins always come visiting if you leave it until the last minute.

To transfer your file to elsewhere, be it a printers or the web, it is much better to output your report as a pdf because word, on different computers with different printers, can sometimes change the pagination. The Adobe software itself is very expensive but there are lots of shareware programs that will allow you to print to pdf. If you haven't had it printed up professionally then you might like to consider a very cheap option to make it look much more important, just get some report covers (the ones with a clear front and a colored back) from you local stationers. They cost pennies but can transform ordinary paper into a much more important looking document.

Make sure you get 10% more copies than you think you'll need as there are usually extra ones needed.

# Distribution

This is your final part of the process. You might like to consider doing a two stage distribution, firstly to whoever asked for the report and then to everyone else.

It is often useful to have a circulation list to make sure that everyone who needs the report gets a copy. You may want to include the distribution list within your report, especially if the report is confidential, so that people are aware of who else will be able to read the report. If you're really proud of your report (and why not, it should be excellent) then consider who else could you send it to that might help your career?

# Celebration

Remember to celebrate your success. You've written the best report you could and hopefully your readers are busy extracting the information into their brains. Keep a copy of it for your own continuing professional development portfolio (CPD).

# Checklist for the stages of writing the report

The following is a checklist for all of the stages of writing a report.

| | Activity | Your notes |
|---|---|---|
| ❑ | Report is requested | |
| ❑ | Going through the 8 questions | |
| ❑ | Setting the objective | |
| ❑ | Checking the objective | |
| ❑ | Design the cover sheet | |
| ❑ | Brain storm the content | |

# CHECKLIST FOR WRITING A REPORT

| | Activity | Your notes |
|---|---|---|
| ❏ | Identify the resources you require | |
| ❏ | Locating the resources | |
| ❏ | Reading the resources | |
| ❏ | Decide what type of report it is | |
| ❏ | Choose your section headings | |
| ❏ | Organize the order of your sections | |
| ❏ | Decide on your formatting styles for your headings, captions and main text | |
| ❏ | Write each section | |
| ❏ | Add further reading/references | |

| | Activity | Your notes |
|---|---|---|
| ❏ | Edit the text, make it as concise as possible and making sure that it matches your objective | |
| ❏ | Write and edit the executive summary | |
| ❏ | Add in the contents, any lists of figures or tables and the index | |
| ❏ | Proofread the report, make sure you look at the title, headers, diagrams, captions and footers, appendices and further reading as well as the main text | |
| ❏ | Proof the layout of the report | |
| ❏ | Final check through | |
| ❏ | Printing/publishing | |
| ❏ | Distribution | |
| ❏ | Celebration | |

# Action Plan

Having worked through the book, have a think about either actions to do with writing your report or actions for further training. Be very specific about each one and give them a date so that you can tick them off as you achieve them.

| Action | Date to be achieved |
| --- | --- |
| | |
| | |
| | |
| | |
| | |
| | |
| | |
| | |
| | |
| | |
| | |
| | |
| | |
| | |

# Final Thoughts

I hope that you've found this book a helpful introduction to report writing. If you liked the planning section then you can actually apply all of it to planning a presentation too. Remember there are lots of free resources on the website, www.uolearn.com.

If you particularly like the style of the book, please do visit the www.uolearn.com website to see what else we offer (we have minute taking, speed reading and speed writing amongst other topics). If you would like a trainer for report writing (or speed reading, improve your memory and creative problem solving) then please do contact me directly (trainer@inspirachange.co.uk).

Now, I'm off to write that executive summary - the final part of the book I have to finish, yippee!

Best wishes and good luck with your career,

# Further resources and reading

## Websites:

**Picture libraries**
www.fotolia.com
www.istockphoto.com
www.shutterstock.com

**Finding acronyms**
www.acronymfinder.com

**Plain English examples**
www.plainenglish.co.uk

**Proofreading checklist**
www.writing.umn.edu/tww/grammar/proofreading.pdf

**Harvard referencing system**
www.libweb.anglia.ac.uk/referencing/files/Harvard_referencing.pdf

**Thesaurus and dictionary**
www.thesaurus.com/

**Resources for go with this book (including the workbook and section on grammar)**
www.uolearn.com

## Time Management:

**Get Everything Done and Still have Time to Play**
Mark Forster, 978-0340746202

## Report Writing

**Report Writing- A practical guide to effective report writing presented in report form**
Gordon Wainwright, 978-0946679010

**Successful Report Writing in a Week**
Katherine Heritage, 978-0340711989

**The Right Report**
Alan Barker, 978-0852909959

**Writing a Report, How to prepare, write and present really effective reports**
John Bowen, 978-1845282936

## Business Writing Skills

**Successful Business Writing,** Heather Baker, 978-1849370110

**Writing Skills, Preparing to write, writing essential, sample letters and reports**, Linguarama, 978-3526511748

**The Pyramid Principle, Present your thinking so clearly that ideas jump off the page and into the reader's mind**
Barbara Minto, 978-0273710516

**Read This! Business Writing that works**
Robert Gentle, 978-0273656500

## Grammar and Writing Skills

**Write Right: A Desktop Digest of Punctuation, Grammar and Style**, Jan Venolia, 978-1580083287

**English Grammar in Use.** Raymond Murphy, 978-0521537629

**Improving your Written English. How to ensure that your grammar, punctuation and spelling are up to scratch.**
Marion Field, 978-1857033588

**The Complete Idiot's Guide to Grammar and Style**
Laurie Rozakis, 978-1592571154

**Proofreading, Plain and Simple**
Debra Hart May, 978-1564142917

**The Elements of Style**, William Strunk, 978-1599869339

**Notes:**

# Universe of Learning Books

"The purpose of learning is growth, and our minds, unlike our bodies, can continue growing as we continue to live." Mortimer Adler

# About the publishers

Universe of Learning Limited is a small publisher based in the UK with production in England, Australia and America. Our authors are all experienced trainers or teachers who have taught their skills for many years. We are actively seeking qualified authors and if you visit the authors section on www.uolearn.com you can find out how to apply.

If you are interested in any of our current authors (including Margaret Greenhall) coming to speak at your event please do visit their own websites (to contact Margaret please visit her website www.inspirachange.co.uk, trainer@inspirachange.co.uk) or email them through the author section of the UoLearn site.

If you would like to purchase larger numbers of books then please do contact us (sales@uolearn.com). We give discounts from 5 books upwards.  For larger volumes we can also quote for changes to the cover to accommodate your company logo and to the interior to brand it for your company.

All our books are written by teachers, trainers or people well experienced in their roles and our goal is to help people develop their skills with a well structured range of exercises.

If you have any feedback about this book or other topics that you'd like to see us cover please do contact us at support@uolearn.com.

To buy the printed books please order from your favorite bookshop, including Amazon, Waterstones and Barnes and Noble. For ebooks please visit www.uolearn.com.

Keep Learning!

# Speed Writing

## Speedwriting for faster note taking and dictation

ISBN  978-1-84937-011-0, from www.UoLearn.com
Easy exercises to learn faster writing in just 6 hours.

- ✓ "I will use this system all the time."
- ✓ "Your system is so easy to learn and use."

# Developing Your Influencing Skills

ISBN: 978-1-84937-022-6 from www.UoLearn.com

- ✓ Decide what your influencing goals are
- ✓ Find ways to increase your credibility rating
- ✓ Develop stronger and more trusting relationships
- ✓ Inspire others to follow your lead
- ✓ Become a more influential communicator

# Coaching Skills Training Course

## Business and life coaching techniques for improving performance

ISBN: 978-1-84937-019-6, from www.UoLearn.com

- ✓ An easy to follow 5 step model
- ✓ Learn to both self-coach and coach others
- ✓ Over 25 ready to use ideas
- ✓ Goal setting tools to help achieve ambitions

A toolbox of ideas to help you become a great coach

# Successful Business Writing

## How to write excellent and persuasive communications

ISBN  978-1-84937-074-5, from www.uolearn.com

- ✓ Think about the purpose of the communication
- ✓ Create successful text for emails, letters, minutes, reports, brochures, websites, and social media
- ✓ Write effective communications to persuade people

# Successful Minute Taking

## How to prepare, write and organize agendas and minutes of meetings

ISBN 978-1-84937-076-9, from www.UoLearn.com

✓ Becoming more confident in your role
✓ A checklist of what to do
✓ Learn what to include in minutes
✓ How to work well with your chairperson

# Practical and Effective Performance Management

ISBN: 978-1-84937-079-0, from www.uolearn.com

✓ Five key ideas to understanding performance
✓ A clear four step model
✓ A large, wide ranging choice of tools
✓ Practical exercises and action planning for managers

A toolbox of ideas to help you become a better leader.

# Stress Management

## Exercises and techniques to manage stress

ISBN: 978-1-84937-024-0, from www.uolearn.com

✓ Become proactive in managing your stress
✓ How to become more positive about your life

An easy 4 step model to lasting change

# Developing your assertive communication skills

ISBN: 978-1-84937-082-0, Order at www.uolearn.com

✓ Decide what you want and communicate it effectively
✓ Develop your confidence
✓ Step by step instructions and worked examples to achieve the results you need